アイアム・ムービースター
00A
0A

KANOSK

OOKING!

KANOCK'S
PLAY
BIG MAC
UNION MADE
SANFORIZED

Y'S mart

BIG MAC
UNION MADE

BIG MAC
UNION MADE

BIG MAC
UNION MADE

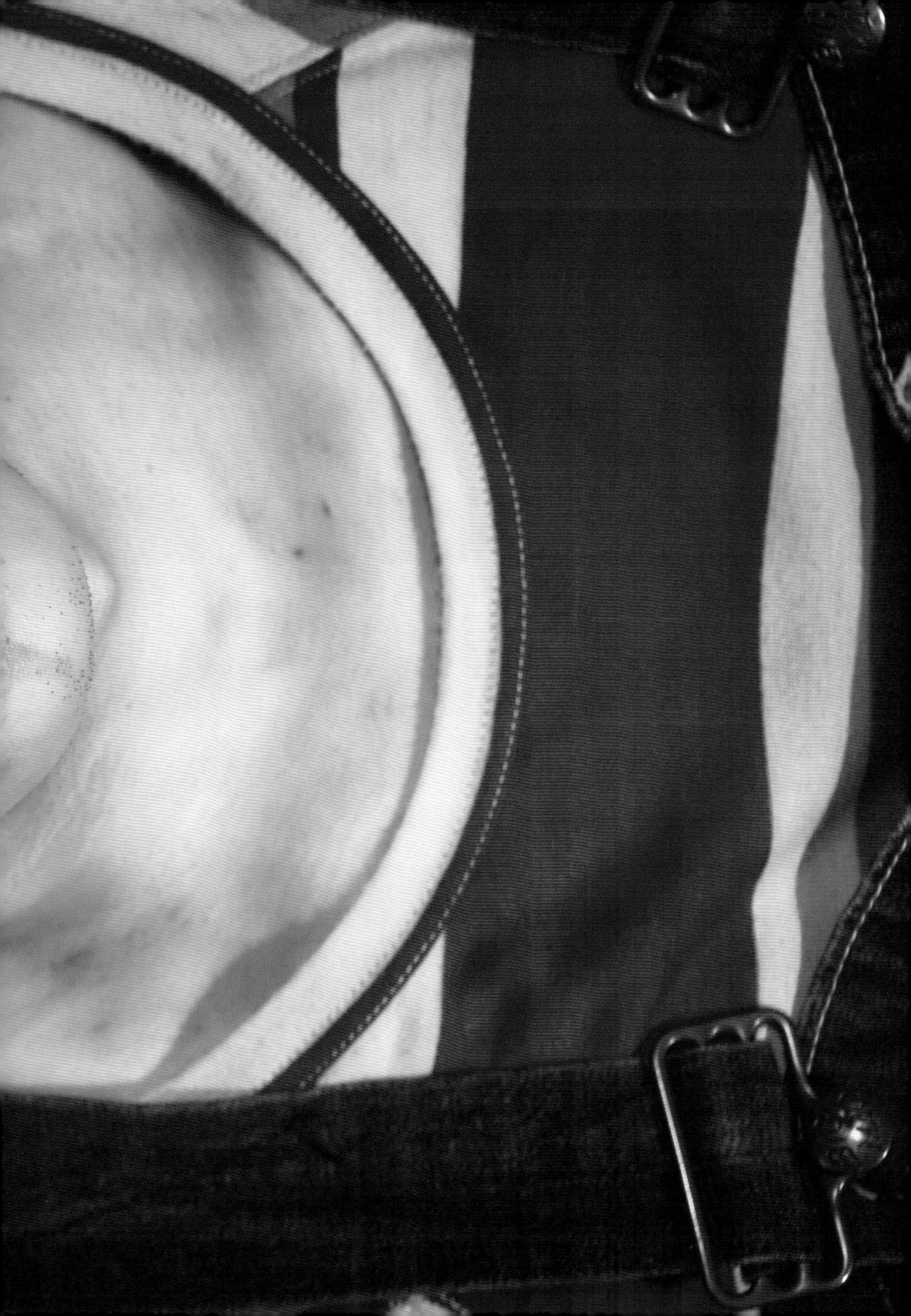

BIG MAC

BIG MAC

BIG MAC

KANOCK TO THE FUTURE

Columbia

Tokai Custom Edition

THE KANOMINATOR

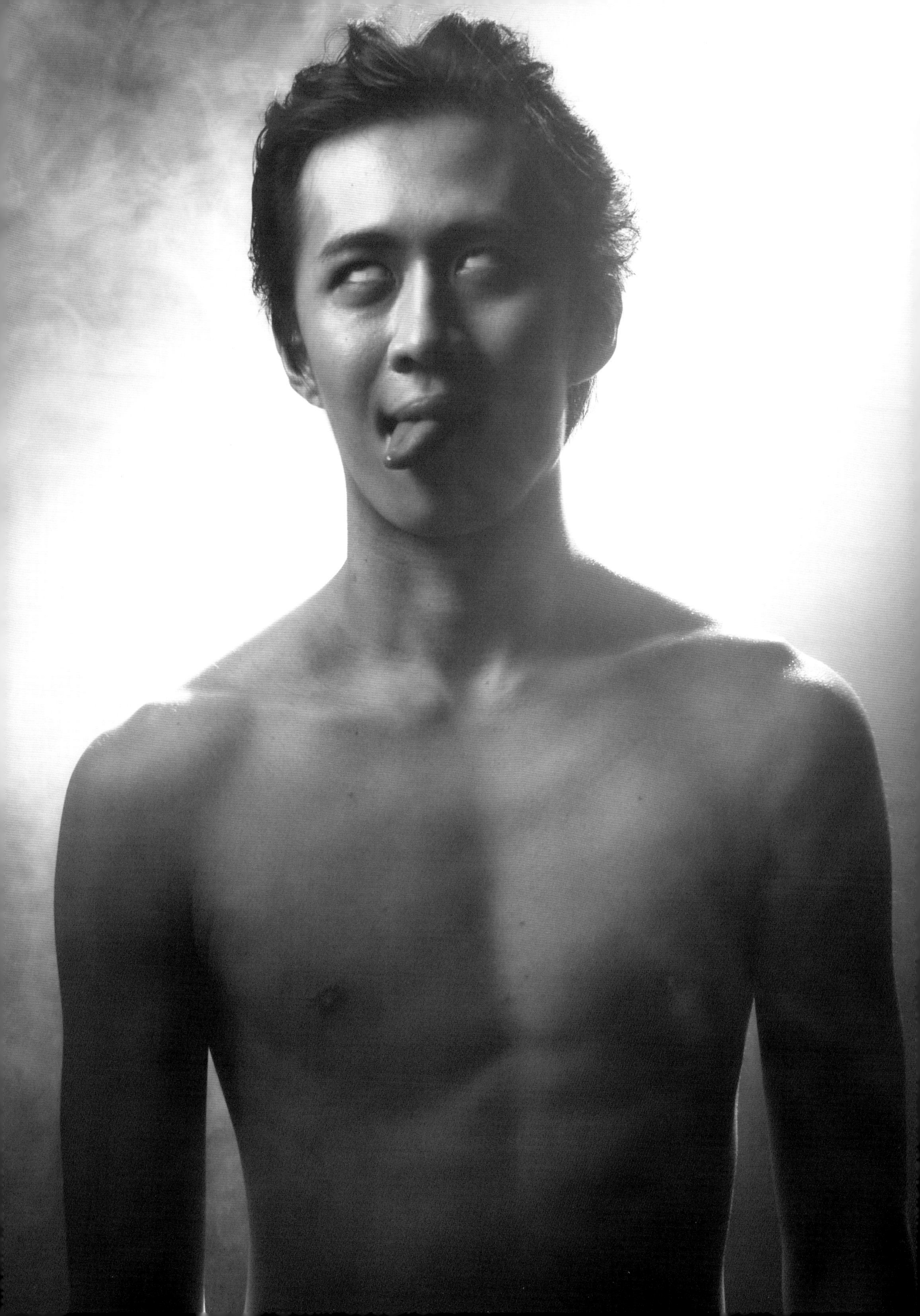

KANO AND THE CHOCOLATE FACTORY

TICKET FRONT

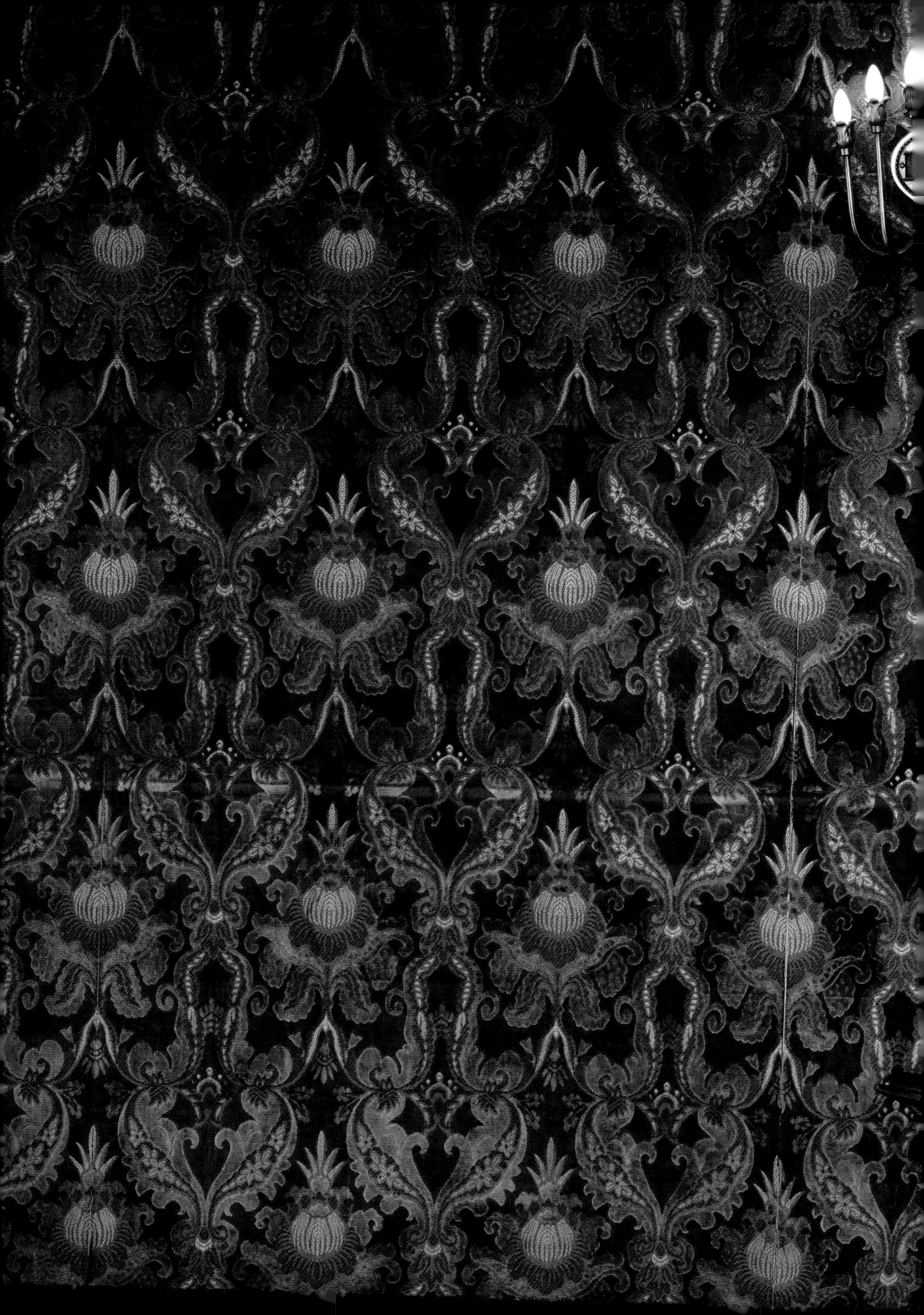

THE KANOCKIES

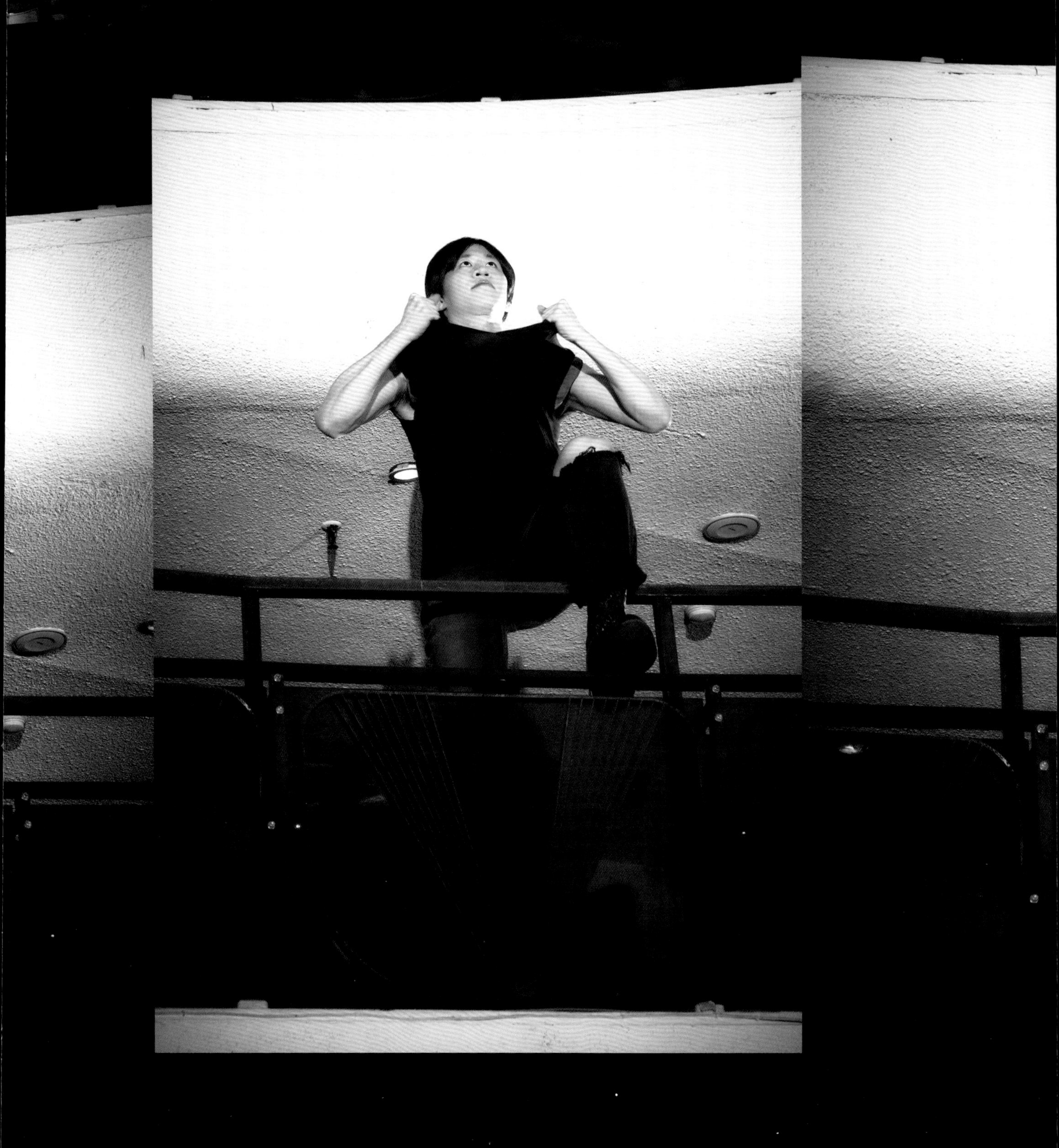

ORGANIC

ORGANIC
ice cream

アイアム・ムービースター

2021年11月24日　初版第一刷発行

Model　カノックスター
Photographer　小野寺廣信（Boulego）
Stylist　甲斐修平
Hair & Make　双木昭夫（クララシステム）

協力　GROVE株式会社

Special Thanks　ぴっちょりーな☆

Transworld Japan Inc.
Produce　斉藤弘光
Designer　山根悠介
Sales　原田聖也

発行者　佐野 裕
発行所　発行所／トランスワールドジャパン株式会社
〒150-0001 東京都渋谷区神宮前6-25-8 神宮前コーポラス
Tel：03-5778-8599　Fax：03-5778-8590

印刷・製本　株式会社グラフィック

ISBN 978-4-86256-327-9
2021 Printed in Japan